All About Us

by Allison K. Lim

For many years, people have used trains to travel. Railroads and train stations were built all over the country. People could travel all the way from Virginia to California!

Train station in Richmond, Virginia

Train stations have always been busy places. People waited in line to buy tickets.

This woman pushed her luggage. She gave her friend a push, too.

Trains moved in and out of stations. They pulled the cars along the tracks. Engines huffed and puffed. Whistles on the trains sounded.

Before the train left the station, the conductor stepped out.

He called out in a loud voice, "All aboard!"

People were in motion. Everyone tried to board the train at once. They pushed and pulled their luggage. They loaded it onto the train.

People found seats on the train. They pulled out their tickets and showed them to the conductor. Then their train ride began!